HI-TECH JOBS WITHOUT COLLEGE

BE A COMPUTER PROGRAMMER

by Tammy Gagne

BrightPoint Press

San Diego, CA

BrightPoint Press

an imprint of ReferencePoint Press, Inc.
Printed in the United States

For more information, contact:
BrightPoint Press
PO Box 27779
San Diego, CA 92198
www.BrightPointPress.com

LIBRARY OF CONGRESS CATALOGING-IN-PUBLICATION DATA

Name: Gagne, Tammy, author.
Title: Be a computer programmer / by Tammy Gagne.
Description: San Diego, CA: ReferencePoint Press, 2026 | Series: Hi-tech jobs without college | Includes bibliographical references and index. | Audience: Grades 7–9
Identifiers: ISBN 9781678212605 (hardcover) | ISBN 9781678212612 (eBook)
The complete Library of Congress record is available at www.loc.gov.

CONTENTS

AT A GLANCE

- Programmers create software using computer code. This is the set of instructions that tells a computer how to carry out specific tasks.

- Many programmers learn to program on their own or with the help of online resources. They do not need a college degree.

- Programmers usually know more than one programming language. They use programming languages to write code.

- Computer programmers can earn certifications, which show they are skilled in a specific language or programming platform.

- A programmer works with team members to solve programming problems.

- Programmers work at an office or remotely from almost anywhere.
- The field of computer programming is always changing. As technology advances, programmers need to keep up with it.
- Computers are now capable of writing code through artificial intelligence (AI). Although some programmers are worried this could put them out of work, others feel strongly that human programmers will still be needed for a long time.

REWRITING THE SCRIPT

Madison Kanna went to college after high school. But she soon realized it wasn't right for her. She wasn't sure what she wanted to do. Madison worried about debt. She didn't want to pay for classes she didn't need. She wanted to learn practical skills. Madison decided to leave college. She needed to think about what she wanted to do. Madison wanted to find a job she loved.

Computer programmers may work together on teams to complete big projects.

Madison had an older sister. Randall was a computer programmer. She lived in San Francisco, California. Madison visited her. Randall was working on a music application. Coworkers were helping, too. Madison was drawn to programming right away. She says, "I just thought it was so magical that they were building something from scratch and that millions of people were going to use this application."[1]

Madison's sister used code to make the app. The code appeared on her computer screen. Madison didn't understand it. But she knew she could learn. She taught herself how to code. It took her about a year. She liked it. Programming was more creative than she thought. People develop their own styles of code.

After learning to code, Madison needed experience. She asked companies about internships. One company took her on. Later, they hired her as a programmer. This experience led to new job offers. Madison eventually became a senior software engineer. She worked for Walmart.

There are many ways programmers can learn to write computer code, including through videos and online tutorials.

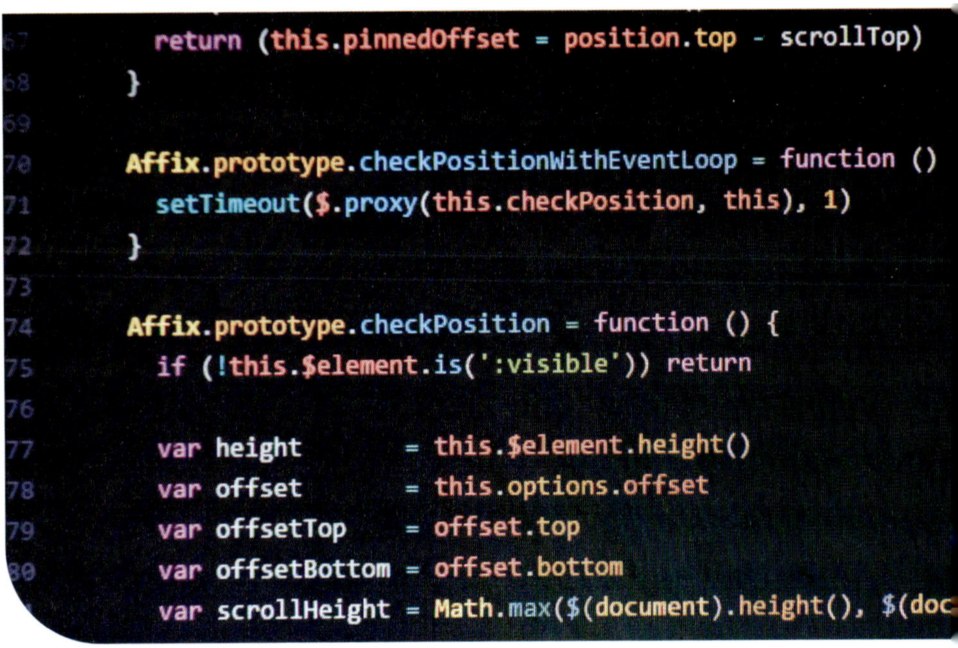

```
    return (this.pinnedOffset = position.top - scrollTop)
  }

  Affix.prototype.checkPositionWithEventLoop = function ()
    setTimeout($.proxy(this.checkPosition, this), 1)
  }

  Affix.prototype.checkPosition = function () {
    if (!this.$element.is(':visible')) return

    var height       = this.$element.height()
    var offset       = this.options.offset
    var offsetTop    = offset.top
    var offsetBottom = offset.bottom
    var scrollHeight = Math.max($(document).height(), $(doc
```

Computer programming involves problem-solving and attention to detail.

COMPUTER PROGRAMMING

Computer programmers use computer code. This is a set of instructions that tells a computer how to do specific tasks. Programmers write, test, and update code. Code is written in programming languages.

There are thousands of these languages. Without programmers, most of the technology people use every day could not exist.

Some programmers have college degrees. They may study computer science. Others study software engineering.

But not all programmers go to college. Some are like Madison. They are self-taught. Others attend workshops. Boot camps are an option, too. These are short training programs. There are many ways people can become programmers.

CHAPTER ONE

EXPLORING COMPUTER PROGRAMMING

Programmers do important work. Billions of people use computer technology each day. They use laptops, tablets, and smartphones. These devices have software. It helps people learn and work. It also helps them communicate. Programmers build that software. Writer Malik Ans explains, "We are living in a world where technology is at the heart of our daily lives. From communicating with friends to . . . using GPS navigation,

The average person spends 6 hours and 40 minutes on tech devices each day.

Binary code can be used to control circuits. The ones turn on the circuits, while the zeroes turn them off.

we rely heavily on various software and hardware applications that are powered by programming."[2]

Computers understand machine language. This language is made of ones and zeros. Data made of ones and zeroes is called binary. It's hard to write in machine language. So programmers created programming languages. These languages use human language. They also use **syntax**.

A set of instructions written in a programming language is called code. Code must be translated for a computer to understand it. A program called a compiler does this work. It translates

A Programming Portfolio

Artists keep a portfolio when looking for a job. A portfolio includes examples of their best work. Computer programmers may also use portfolios. They should have examples of their work. These examples should show technical skills. They should also show creativity and problem-solving abilities.

code from a programming language to machine language.

Programming languages include markup languages. There are also scripting languages. Each type of language plays a role in programming.

Programmers use different computer languages for different jobs. Some work well for general use. Others work best for mobile apps or gaming.

The Bureau of Labor Statistics provides information about jobs. It states that most programmers know several programming languages. Many programming jobs require the use of more than one language.

COMMON LANGUAGES

Programming languages tell a computer what to do. They also explain how to do it. Programming languages are used to make software. This includes mobile apps and video games. Java, Python, and C++ are common programming languages.

Scripting languages do specific tasks. These tasks are called scripts. They help different parts of a program talk with each other. These languages are used to make websites. JavaScript, Perl, and Ruby are scripting languages.

JavaScript and Java sound alike. But they are completely different. In 1995, Brendan Eich created LiveScript. It took him just 10 days. LiveScript is a scripting language. It was first used for Netscape. This was an early web browser. The company behind Netscape changed the name from LiveScript to JavaScript. It wanted people to see that JavaScript works with Java. JavaScript is still popular. In 2024, 64.6 percent of programmers used it. PHP is another scripting language.

Alex Zelinsky uses JavaScript to make websites. She states, "JavaScript is a great language to start with. It has a relatively simple syntax, and there are

HTML and CSS are not considered true programming languages. They help format content rather than directing the computer to perform tasks.

a ton of resources out there to help you get started."[3]

Markup languages do not tell a computer what to do. They format text. This affects how a program looks. They also arrange elements on a page. One is Hypertext Markup Language (HTML). It is often used

Future computer programmers can save money by taking only the classes they need rather than getting a full college degree.

on websites. Cascading Style Sheets (CSS) is used with HTML. HTML creates the content and structure of a web page. CSS changes the way the web page looks.

All of these languages do different things in programming. It may help to think of a program like a house. Programming languages are tools. Programmers use them to build the house. Scripting languages are like wiring. They help different parts of the house work. Markup languages are like decorations. They help the house look nice.

NO DEGREE NEEDED

Some programmers have college degrees. But many tech companies don't require a degree. They look at a person's skills. These skills can come from degree programs. But they can also come from other paths.

Companies look for people with certain backgrounds. These include science, technology, engineering, and math (STEM). People who like STEM subjects often do well in this career. Jenny Wem is a computer programmer. She explains that it is not an **impediment** if a programmer does not have a degree in a STEM subject. But she thinks those who have been involved in STEM from a young age are more likely to go into computer programming.

A degree might help a programmer find work. But curiosity can be even more important. Programming is always changing. The best programmers master basic skills. But they also keep learning. This is key to being a good computer programmer.

Successful programmers practice their skills and continue to look for new ways to solve problems.

CHAPTER TWO

TRAINING FOR COMPUTER PROGRAMMING

Good programmers have certain skills and traits. Strong math and writing skills help. Knowing algebra can make it easier to write code. Logic can help, too. These skills are used to study **algorithms** as well.

Being able to write clearly is important, too. This allows programmers to explain their code to team members. Programmers should be comfortable working alone

The skills used when solving math equations can be helpful when learning to write computer code.

Many people recommend Python for beginning programmers because it is easy to use and understand.

or with a team. They should also enjoy research. This is how they learn new ways to do things. Programmers must use what they have learned to solve problems.

Beginners should think about what type of programming they want to do. This affects which languages they should learn. For example, JavaScript and PHP are the best languages for web development. Swift is a programming language. It was created by Apple. Swift is used to create apps for Apple's computers and smartphones. Python is a programming language. But it is also used for scripting. It's easy to learn. It's useful in many industries.

People can take classes to learn languages. These may be in a classroom or online. Classes may be offered at colleges or community centers. Online platforms offer programming classes, too. Many basic classes are free. More advanced classes may require a fee.

INDEPENDENT LEARNING

Many programmers are self-taught. Books can be a good option. This is especially true for people who like to learn on their own. Books often go into greater detail than online courses do. Finding the most recent books is important. Programming is like all technologies. It is always changing.

Other options are videos or podcasts. Watching YouTube videos can be a great way to learn about programming. A person who is struggling with a concept can search for a video about it.

Troubleshooting is a big part of programming. Videos posted by experienced programmers can be helpful. They may show solutions to common problems.

Watching online videos can be an effective way to learn how to code.

Many people learn about computer programming in classrooms. It can be helpful to bounce ideas off classmates. Other people like finding answers on their own. Practice is key when learning a programming language. Dennis Ritchie invented C. This is one of the most used

Web Development

Frontend developers focus on user interface (UI) or user experience (UX). This includes the buttons, sliders, and dropdown menus users see and use. Frontend uses HTML and JavaScript. Backend developers work behind the scenes. They make sure websites, apps, and games work properly. Backend includes databases, servers, and security. These programmers use Golang and SQL. Full stack developers can work in both roles.

In-person or online computer programming classes may be the best ways to learn for some students.

programming languages. He says, "The only way to learn a new programming language is by writing programs in it."[4]

CERTIFICATIONS AND BOOT CAMPS

Not everyone likes learning on their own. And many people don't want to go to college. Those who want to learn programming have other options.

These include certification programs and boot camps. Programming certificates show that someone knows a language or programming platform. A certificate tells employers that a person's skills match **industry standards** for a programming job. A certification can make it easier for a person to get a job.

Certificate programs are offered by universities. Technology companies offer them, too. They last about a year. Oracle is a technology company. It offers the Java SE 8 Programmer certificate. This certificate takes 120 hours to complete. People pay a fee. They must pass an exam at the end of the course. This certification can help people find entry-level jobs where Java is used.

Oracle offers a number of certification paths, including several for Java SE.

The Microsoft Certified: Azure Developer Associate certification is another option. Azure is a **cloud computing** platform. This program takes between 40 and 50 hours to complete. People must have 1 to 2 years of experience with Microsoft Azure to enroll.

People learn how to design, build, and test cloud applications on Microsoft Azure.

Boot camps train people for entry-level programming jobs. Boot camps are usually short. But they are difficult. Some camps

Microsoft Azure is a cloud computing program that provides storage, networking, and databases.

take 12 weeks. Others take up to 7 months. Camps cover practical skills in a specific area. Coding boot camps may focus on JavaScript, HTML, and CSS. Students also learn popular programming languages. Python, Java, and C++ courses are standard. Boot camps teach key skills. But the quality of a course can vary. Unlike certification, boot camps do not offer official recognition of a skill.

There are many options for learning how to program. Several resources are free. Learners should make sure a paid course provides quality content. They can look up reviews of a course. They can try to contact people who have taken it.

CHAPTER THREE

WORKING IN COMPUTER PROGRAMMING

Many people think programmers just write code. This is a big part of a programmer's day. But there are other important tasks, too. Programmers spend time getting organized. They usually start by checking their emails. They also check their calendar. This helps them set goals for the day.

Some programmers work on a team. They go to a stand-up meeting

Some programmers begin each workday with a stand-up meeting. Others may update team members through messaging apps such as Slack.

each morning. This means people often stand for these meetings. It keeps the meetings short. Programmers share their plans for the day. They talk about what they did the day before. They discuss challenges, too.

After morning meetings, programmers work. They may write code. Or they may

Waterfall vs. Agile

There are two different types of workflow in programming. They are called Waterfall and Agile. Waterfall means the programmer gets all the project details at the beginning. They then work until the project is done. Agile means the programmer works on small parts of a project at a time. Different phases take shape along the way. Most job postings list one of these workflow styles.

Having blocks of focused time allows programmers to complete projects faster.

test code. They look for errors, called bugs. This is called debugging code. Many programmers prefer not to be disturbed during this time. Vitaly Tev works for CodeMonkey. This online platform teaches coding to kids. He explains, "Programming requires a high level of concentration. So many developers prefer uninterrupted blocks of time, often referred to as 'deep work.'"[5]

During lunch, programmers eat and talk with coworkers. Some take a walk. Others go to a gym. Programming requires long periods of focus. Breaks are important. They help avoid **burnout**. They allow

Teaming up with other programmers to write code can be more efficient. It can also help people produce code with fewer errors.

programmers to clear their minds. This helps them reset for the afternoon.

After lunch, many programmers work on team-focused tasks. Programmers are part of tech teams. Together they create apps, websites, and other software. They work together with designers, testers, and project managers.

They do code reviews of other programmers' work. This helps them catch mistakes. Reviews also help newer programmers learn. Programmers discuss problems they must solve. Team members can offer solutions. Programmers may continue their own coding and debugging. Some programmers make notes at the end of a day. This helps them remember where they left off.

HOURS AND WORK ENVIRONMENT

Many computer programmers work 40-hour weeks. Programmers who work for a company usually work Monday through Friday. They have weekends off.

Self-employed programmers set their own schedules. They may work nights or weekends if needed. Other times, they may take mornings or some weekdays off. Computer programmers often work on big projects. Sometimes they need to work extra hours. This helps them finish jobs on time. They may work through the lunch hour.

Programmers who work for tech companies may work in an office. Some work remotely. This means they work at home or from other locations. They can

According to a 2025 survey, 50 percent of computer programmers spent some days in an office and worked remotely the rest of the week.

use a laptop to work almost anywhere. They may work in a busy coffee shop. Others may work at a park. However, some companies require a secured internet connection for working remotely.

Chapi Menge is a backend programmer. He works remotely. Menge says working from home has pros and cons. Remote work gives workers more freedom. Menge states, "I will be honest here, I am not a

Co-op office spaces offer people places to work away from home. This is an option for those who work remotely.

morning person or early bird. So I love to wake up late and start my day late."[6] But he misses joking with coworkers. He also misses eating lunch with other people. And he thinks some topics are easier to discuss in person. Some programmers work partly at home and partly in an office. This is called hybrid work. It can offer the best of both types of work.

PROGRAMMING TOOLS

Programmers need the right hardware and software. This helps them work **efficiently**. Every programmer must have a computer. This is hardware. Some programmers prefer to work on a desktop computer. This larger system is usually more powerful than a laptop. This means it runs faster. It may also handle more complex programs. But laptops are easy to move from one spot to another. This can make working with team members easier. On both laptops and desktops, many programmers like to use multiple monitors.

Programmers use several types of software. One is a code editor. Examples are Visual Studio Code or Sublime Text. Code editors are where programmers write their code. Many have built-in debuggers. Programmers may also use a separate debugging program. They need a compiler or interpreter as well. These work like language translators. They allow the computer to understand and run the code the programmer writes.

Finally, programmers must have a version control system. This software allows programmers to work on the same project as other programmers. The software safely combines each person's work into the project. It prevents issues such as saving over someone else's work.

Internet access is not needed to write code. But it can be helpful to programmers. The internet helps with research. Programmers can look up

MOST COMMONLY USED PROGRAMMING LANGUAGES

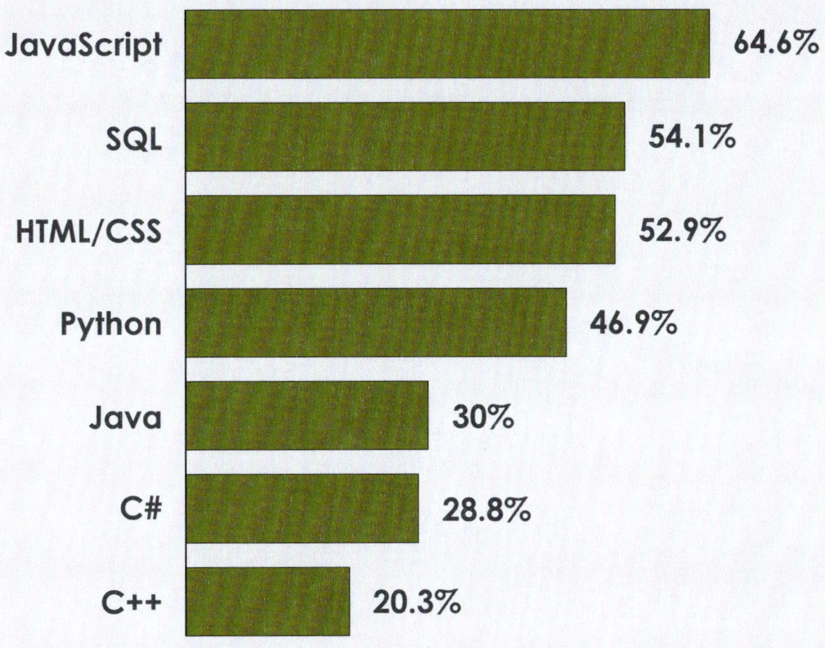

Source: "Most Popular Technologies: Programming, Scripting, and Markup Languages," Stack Overflow, *2024. https://survey.stackoverflow.co.*

Researchers asked computer programmers what language they use to write code. Programmers use many programming languages, so many picked more than one in this survey.

ideas for troubleshooting. They can also contact other programmers, managers, or team members.

LOOKING AHEAD

Technology is always advancing. A tool that was new just a year ago may be outdated now. Even products that have been around for years are always being updated. This means tech jobs are always changing, too.

Programmers must keep up with new technology. This means they are always learning. New languages may solve problems that are common in

Programmers can learn new skills through discussions with peers. These discussions can happen in office settings or in online forums.

older languages. For example, Python uses fewer lines of code than Java to do the same things. Swift can avoid many coding errors that may happen when using C-based languages. And Rust was designed to prevent security issues that are common in earlier languages.

Keeping up-to-date on the latest technologies allows programmers to work more efficiently. This gives them time to tackle more difficult projects.

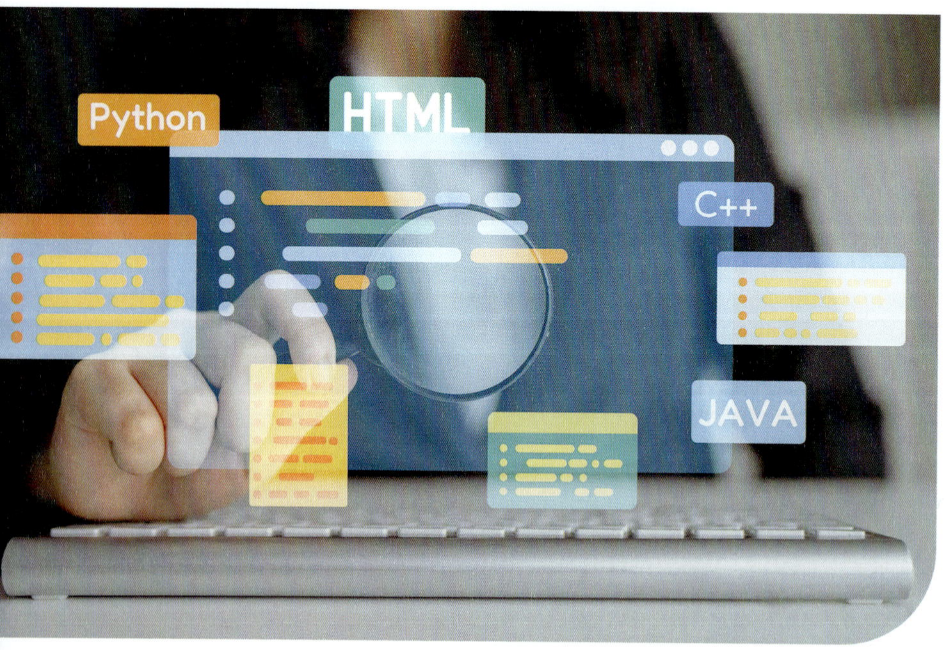

Programmers must also know the latest programming tools and trends. Nimrod Kramer is a programmer. He explains:

> *If you don't make an effort to learn new stuff, it could really hurt your career. As new tools and ways of doing things become popular, your current skills might not be enough anymore. Developers who don't learn regularly might find it hard to get or keep a good job.*[7]

THE RISE OF AI

In the early 2020s, artificial intelligence (AI) became a big part of computer technology. AI can now perform many tasks that once required humans. Computer programs can recognize human faces. They can understand speech. Others can translate one language into another. Still others can do research. Computer programs can even use the information they gather to make decisions.

AI-driven programs can do lengthy tasks. This includes writing repetitive code. AI can also debug code. It can give quick solutions to coding problems. This allows

A Good or Bad Thing?

People often disagree about whether AI is a positive or negative tool. Some people think AI will hurt their careers. Other people are excited about AI. They want AI to improve their jobs. According to Pew Research Center, 32 percent of tech workers think AI will help them. Only 11 percent think it will do more harm than good.

AI-powered virtual assistant apps were among the most popular apps downloaded for smartphones in 2025.

programmers to get more done. Efficiency frees up programmers. This means they can solve more challenging problems. Humans are able to think creatively and define problems. AI is not yet able to do these things.

As AI improves, it will likely perform more tasks in computer programming. ChatGPT

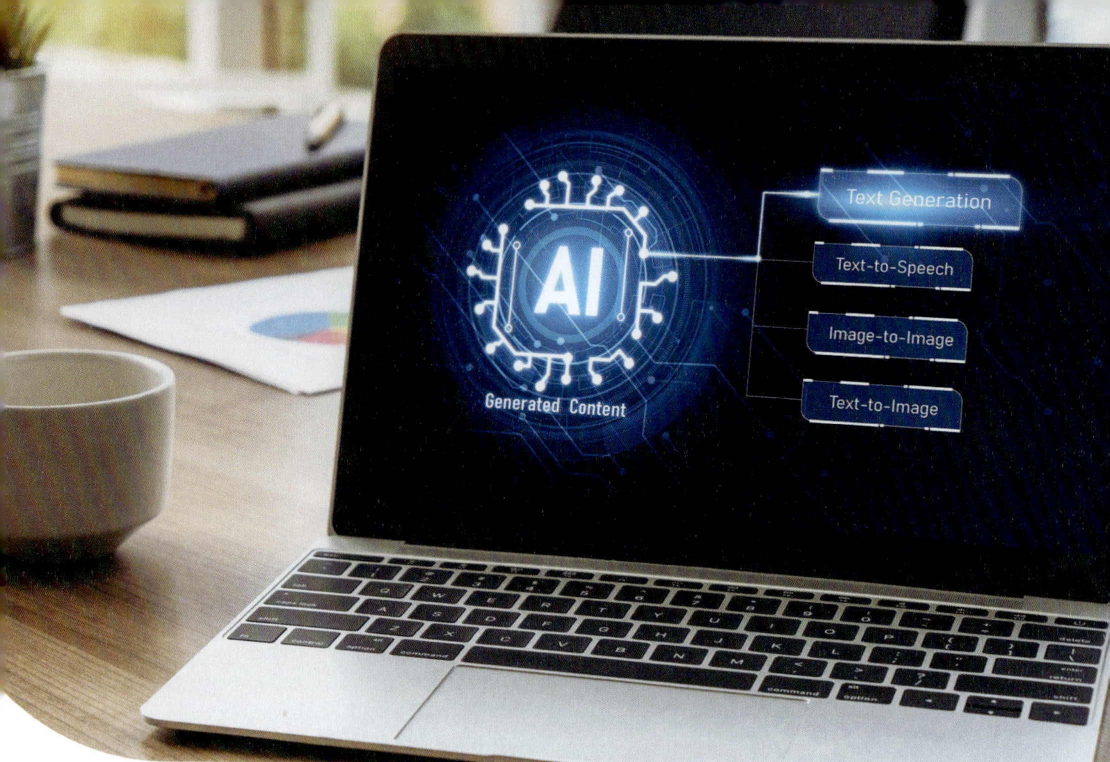

AI will change the way computer programmers work in the years ahead.

is a popular AI tool. In 2025, it already had the ability to write code. In the future, AI is likely to perform some tasks faster than people. It could replace many entry-level programming jobs.

But code created by AI often has errors. Faulty code doesn't work. Humans have to check the code. This makes more work for programmers. Using AI to write code can also create security problems. It gives others access to data. This can expose a company's private information.

THE CHANGING ROLES OF PROGRAMMERS

Some experts say that AI will actually create more programming jobs. This is partly because of the high demand for software. Kevin Dewalt is a software developer. He states:

> *Programming is not just about writing code. The essence of creating software lies in creativity, defining problems, breaking them down, troubleshooting, and effective communication. These are . . . human skills that AI is yet to* ***replicate.***[8]

Future programmers will need to know how to work with AI. This will help them work efficiently. But it's important to work on other skills, too. Keeping up with changes in programming languages is key. Programmers should also keep improving other skills. This includes problem solving.

AI can help programmers get more done, but work created with AI also needs to be checked for errors.

In the future, continuing to learn about the latest developments in the field will be important for computer programmers.

No matter a person's training, they will need these skills. This knowledge will help them build a career in the changing world of computer programming.

GLOSSARY

algorithms

sets of instructions or rules for solving problems

burnout

a condition in which too much stress causes a person to lose interest and focus

cloud computing

the on-demand availability of computing resources, such as storage, over the internet

efficiently

done quickly and effectively

impediment

a problem that gets in the way of reaching a goal

industry standards

rules and specifications that are followed in a field of work

replicate

to copy exactly

syntax

the proper order of words in a language

troubleshooting

trying different ways to solve a problem

SOURCE NOTES

INTRODUCTION: REWRITING THE SCRIPT

1. Quoted in *Code Newbie*, podcast, season 24, episode 8, "A Model's Journey to Software Development with Madison," June 28, 2023. www.codenewbie.org.

CHAPTER ONE: EXPLORING COMPUTER PROGRAMMING

2. Malik Ans, "What Is the Role of Programming in Our Daily Life," *Medium*, January 10, 2023. https://medium.com.

3. Alix Zelinsky, "Why I Love JavaScript: A Developer's Perspective," *Medium*, July 26, 2022. https://medium.com.

CHAPTER TWO: TRAINING FOR COMPUTER PROGRAMMING

4. Quoted in Claire D. Costa, "Top Programming Languages and Their Uses," *KD Nuggets*, January 19, 2022. www.kdnuggets.com.

CHAPTER THREE: WORKING IN COMPUTER PROGRAMMING

5. Vitaly Tev, "What Does a Day in the Life of a Programmer Look Like?" *CodeMonkey*, January 2, 2025. www.codemonkey.com.

6. Chapi Menge, "Remote vs. Office: Reflecting on My Journey as a Software Engineer," *Chapi Menge's Blogs*, October 17, 2023. https://blog.chapimenge.com.

CHAPTER FOUR: LOOKING AHEAD

7. Nimrod Kramer, "General Programming and Continuous Learning: Staying Updated," *Daily.dev*, April 9, 2024. https://daily.dev.

8. Kevin Dewalt, "AI Won't Replace Programmers," *Medium*, December 13, 2023. https://medium.com.

FOR FURTHER RESEARCH

BOOKS

Kari Cornell, *Be a Web Designer*. Brightpoint Press, 2025.

Tammy Gagne, *Be a Systems Administrator*. Brightpoint Press, 2026.

Andrew Yueh, *JavaScript Coding for Teens: A Beginner's Guide to Developing Websites and Games*. Rockridge Press, 2021.

INTERNET SOURCES

"Computer Programmers," *My Next Move*, n.d. www.mynextmove.org.

"Crash Course Computer Science," *PBS LearningMedia*, 2025. https://pbslearningmedia.org.

"Sam Haupert: Software Developer," *Roadtrip Nation*, 2025. https://roadtripnation.com.

WEBSITES

Computer Science in the Real World

https://code.org/en-US/students/careers-in-computer-science

Computer Science in the Real World is a resource for learning computer science skills. It also includes interviews with people who work in the computer science industry.

International Web Association (IWA)

www.iwanet.org

The International Web Association provides classes and certifications in web-related fields. It is also a source for those looking for jobs.

Teens in AI

www.teensinai.com

Teens in AI is a website that encourages teens to get involved and share ideas and insights on AI topics. It offers courses and workshops to help teens learn about AI.

INDEX

IMAGE CREDITS

Cover: © Gorodenkoff/Shutterstock Images
5: © PeopleImages.com-Yuri A./Shutterstock Images
7: © NDAB Creativity/Shutterstock Images
9: © Cronislaw/Shutterstock Images
10: © Kindamorphic/iStockphoto
13: © BongkarnGraphic/Shutterstock Images
14: © matejmo/iStockphoto
16: © anyaberkut/iStockphoto
19: © Rawpixel.com/Shutterstock Images
20: © Prostock-Studio/Shutterstock Images
23: © PeopleImages.com-Yuri A./Shutterstock Images
25: © Ground Picture/Shutterstock Images
26: © Chay_Tee/Shutterstock Images
29: © Lordn/iStockphoto
31: © Gorodenkoff/Shutterstock Images
33: © Sundry Photography/iStockphoto
34: © bangoland/Shutterstock Images
37: © Jacob Lund/Shutterstock Images
39: © Africa Studio/Shutterstock Images
40: © Gorodenkoff/Shutterstock Images
43: © puhhha/Shutterstock Images
44: © ESB Professional/Shutterstock Images
47: © Red Line Editorial
49: © NDAB Creativity/Shutterstock Images
50: © Summit Art Creations/Shutterstock Images
53: © Kenneth Cheung/iStockphoto
54: © Summit Art Creations/Shutterstock Images
56: © Anderson P./Shutterstock Images
57: © Gorodenkoff/Shutterstock Images

ABOUT THE AUTHOR

Tammy Gagne is an author and editor with a passion for educational nonfiction. She has written hundreds of books for both adults and young people. Residing in the beautiful state of Maine, she enjoys life with her husband, their son, and two rescue dogs. When not writing or editing, she is often brainstorming her next project. She hopes her books inspire and educate readers of all ages.